# YOUR KNOWLEDGE HAS VALUE

**Dane Marie Lacap et al.**

# Preference of CFAD Students on Different Brands of Shoes

GRIN Publishing

**Bibliographic information published by the German National Library:**

The German National Library lists this publication in the National Bibliography; detailed bibliographic data are available on the Internet at http://dnb.dnb.de .

**Imprint:**

Copyright © 2015 GRIN Verlag GmbH
Print and binding: Books on Demand GmbH, Norderstedt Germany
ISBN: 978-3-656-96426-1

**This book at GRIN:**

http://www.grin.com/en/e-book/299829/preference-of-cfad-students-on-different-brands-of-shoes

**GRIN - Your knowledge has value**

Since its foundation in 1998, GRIN has specialized in publishing academic texts by students, college teachers and other academics as e-book and printed book. The website www.grin.com is an ideal platform for presenting term papers, final papers, scientific essays, dissertations and specialist books.

**Visit us on the internet:**

http://www.grin.com/

http://www.facebook.com/grincom

http://www.twitter.com/grin_com

# PREFERENCE OF CFAD STUDENTS ON DIFFERENT BRANDS OF SHOES

A.Y. 2014 – 2015

_____

A research paper

Submitted to

University of Santo Tomas

_____

In Partial Fulfillment

For Statistics

_____

Submitted by:

Cruz, Patricia Celene                    Lacap, Dane Marie

Guardiano, Clarice Lou                   Lantin, Ivan Lendl

Infante, Russelle Dalbert                Muriel, Lyka

# ACKNOWLEDGEMENT

The researchers would like to express the deepest gratitude to the following that supported and helped to make this research possible. To Mr. Crisencio M. Paner, Statistic professor of College of Fine Arts and Design from University of Santo Tomas, who provided insights and expertise that perfectly assisted the research papers. Who taught and patiently helped the researchers all the way to finish the papers about the perspective of CFAD student in terms of choosing and buying shoes.

To Mrs. JaezamieViray-Ong, Literature professor of College of Fine Arts and Design form University of Santo Tomas, for assistance and correcting the survey questionnaires that greatly improved and helped the researchers to get an accurate result.

The researchers would also like to extent their gratitude for the students of the College of Fine Arts and Design, who answered the survey questionnaires thoroughly, sharing their perspective in choosing and buying shoes to wear in University.

For the moral support that were received from the close friends and suggestions during the course of their research. Thank you.

Above all, utmost appreciation is extended to the All Mighty God for giving the researchers enough knowledge and determination to finish this research.

# ABSTRACT

The purpose of this study is to identify the preferred brands of CFAD students. Thus, the gathered information gives the advantages and disadvantages of the students' preferred brands by rating its comfort, design, durability, affordability, and overall quality. The researchers made an online survey which they sent links to one hundred random CFAD students through Facebook. Likewise, with the researchers being CFAD students themselves thought of the selected brands. However, the study is not limited only to the selected brands but the respondents have the option where they could state the brand that they prefer more over the given selected brands. They would choose their preferred brand for school shoes, sneakers, and rubber shoes. This study would help those who would like to know the preferred shoes Fine Arts students for marketing tactics and the like.

# TABLE OF CONTENTS

# CHAPTER I: INTRODUCTION

## BACKGROUND OF THE STUDY

Shoes have been a good companion in people's everyday lives. It tells and talks about the story of who the person really is: characteristics, hobbies, status, gender, ethnicity, likes, religion, and profession. It became a part of human lives. As time goes by fashion have dictates a lot of element designs on a shoe whether it is flat or heels, leather or rubber. It is made to differ profession and social status. School shoes for students, sleek leather shoes and stiletto heels for business men and women, and sometimes shows elegancy, sneakers for athletes, laid black flats for ladies who are seeking for comfortable footwear. It is made for uniformity of the status society dictates.

As university students, CFAD students can carry their own style compared to other colleges even when they are wearing the required uniform. Likewise, outside of the university they are more knowledgeable on wearing the right clothes, shoes, their brands, and latest trends. Knowing the preferred brands of CFAD students and why they prefer it would identify some advantages and disadvantage. Thus, the gathered information could be used when marketing for fine arts students.

## STATEMENT OF THE PROBLEM

The researchers aim to find out what a customer, specifically CFAD students, look for when buying shoes and their preferred brand. They would like to know the answer to the following questions:

• What are their preferred brands not limited to the following selected brands?

• What are advantages and disadvantages of the preferred brands based on its comfort, design, durability, affordability, and overall quality?

## SIGNIFICANCE OF THE STUDY

This study will benefit the following people:

**The researchers and the respondents**. This would give them an idea on which of the selected brands is most preferred and why are these are most appealing to CFAD students.

**The marketing management**. This study would give them the idea of how Fine Arts students choose brands and the reason why they have chosen it based on the ratings. Thus, this be used when Fine Arts students are chosen as target market.on.

OBJECTIVES

This study aims to find out the preference of CFAD students on different brands of shoes. Thus, analyze what they consider on a brand. The researchers would also like to find out how the following criterion: comfort, design, affordability, durability, and overall quality; affect their buying decision.

SCOPE AND LIMITATION

SCOPE

This study covers the preferences of CFAD students on different shoe brands with selected brands of shoes based on the thought of researchers which are: Gibi, Russ, and Hush Puppies for school shoes; Converse, Vans, and Keds for sneakers; and Nike, Adidas, and Advan for rubber shoes. The study will also cover the factors a person considers when buying shoes and rated by the respondents: price, comfort, design, durability, and its overall quality. The respondents are one hundred CFAD students that are chosen randomly and surveyed through a survey website through sending links by means of Facebook.

LIMITATIONS

The study is not limited to the selected given brands of shoes. The respondents have an option to state their preferred brand over the selected given brands. Thus, this study will only be conducted only on CFAD students with having 40 male students and 60 female students not limited to their program major or year level that will focus on rating the preferred brands' comfort, design, durability, affordability and overall quality.

## CHAPTER 2: RESEARCH METHODOLOGY

## SETTING OF THE STUDY

The study took place through a survey website, SurveyMonkey during the second semester of school year 2014-2015. SurveyMonkey is an online survey development cloud based ("software as a service") company, founded in 1999 by Ryan Finley. SurveyMonkey provides free, customizable surveys, as well as a suite of paid back-end programs that include data analysis, sample selection, bias elimination, and data representation tools. Thus, the links of this website were sent by means of Facebook.

## SUBJECT OF THE STUDY

One hundred students of the College of Fine Arts and Design (CFAD) from University of Santo Tomas were randomly selected to be respondents for the research study to answer one of the researcher's goal of knowing the perspective of CFAD students in terms of choosing and buying shoes: school shoes, sneakers and rubber shoes. The respondents were randomly selected and classified from year level and gender and program major to give fair chances to be part of the study

## PROCEDURE

The first step in the procedure of the research is to identify the problem and set the goals. The researchers agreed that the problem is the preference of CFAD students with regards to different brand of shoes. In this procedure, the researchers reviewed the existing data and formulated research goals and hypotheses.

After this, the researchers selected method and sample. They identified the target population, sample population, determined the sample size of 100 and the type of sampling is random, and determining the method of sampling as mail or online.

For the third step, the researchers constructed a questionnaire. After the questionnaire has been validated, the researches prepared data collection. They gave the survey link form to random students from the College of Fine Arts and Design. When the data collection is done, they conducted the data analysis. The researchers entered the data they gathered, and then reviewed their research goals and hypotheses.

The last part of the research would be formulating the conclusion.

## CHAPTER 3: RESULTS AND DISCUSSION

The researchers conducted a survey on CFAD students to find out their different preferences of shoe brands, with the goal of 100 respondents the researchers approached different year levels and various courses under the program major. The researchers were able to gather 22 (22.22%) respondents from the freshmen, 34 (34.34%) respondents from the sophomores, 20 (20.20%) respondents from the third years and 23 (23.23%) respondents from the fourth years, with a total of 99 (100%) respondents.

| YEAR LEVEL | FREQUENCY | RELATIVE FREQUENCIES | PERCENTAGE |
|------------|-----------|----------------------|------------|
| $1^{st}$ | 22 | 0.22 | 22% |
| $2^{nd}$ | 35 | 0.35 | 35% |
| $3^{rd}$ | 20 | 0.20 | 20% |
| $4^{th}$ | 23 | 0.23 | 23% |
| TOTAL | 100 | 1.00 | 100% |

Table 1.a : Frequency distribution of Year Level

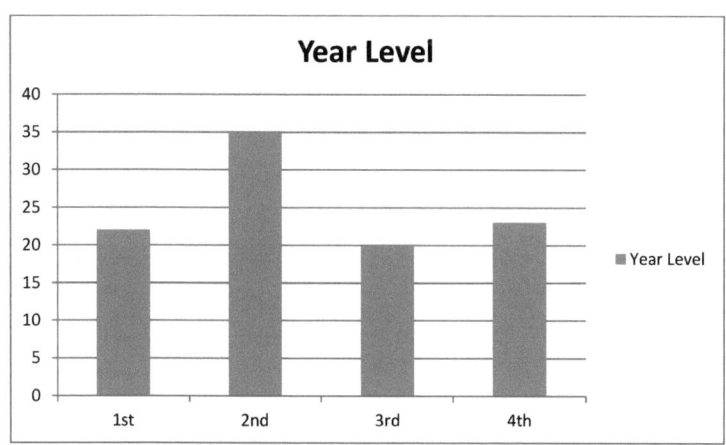

*Table 1.b: Graphical Display of Year Level*

In this survey the researchers were able to gather respondents from different program majors, from the advertising arts there were 63 (65.63%) respondents, industrial design had 11 (11.46%) respondents, and interior design has 8 (8.33%) respondents and painting which had 14 (14.58%) respondents, in an overall total of 96 (100%) respondents. In the gender category, the researchers were able to gather 39 (39.39%) male respondents and 60 (60.61%) female respondents.

| MAJOR | FREQUENCY | RELATIVE FRIQUENCY | PERCENTAGE |
|-------|-----------|--------------------|------------|
| AD | 64 | 0.64 | 64% |
| IND | 12 | 0.12 | 12% |
| ID | 9 | 0.09 | 09% |
| PTG | 15 | 0.15 | 15% |
| TOTAL | 100 | 1.00 | 100% |

*Table 2.a: Frequency distribution of program majors*

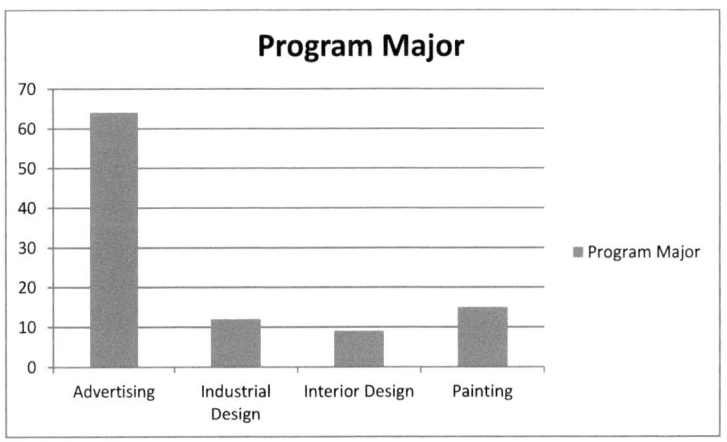

Table 2.b: Graphical display of program majors

| GENDER | FREQUENCY | RELATIVE FREQUENCY | PERCENTAGE |
|--------|-----------|--------------------|------------|
| MALE | 40 | 0.4 | 40% |
| FEMALE | 60 | 0.6 | 60% |
| TOTAL | 100 | 1.0 | 100% |

Table 3.a : Frequency distribution of Gender

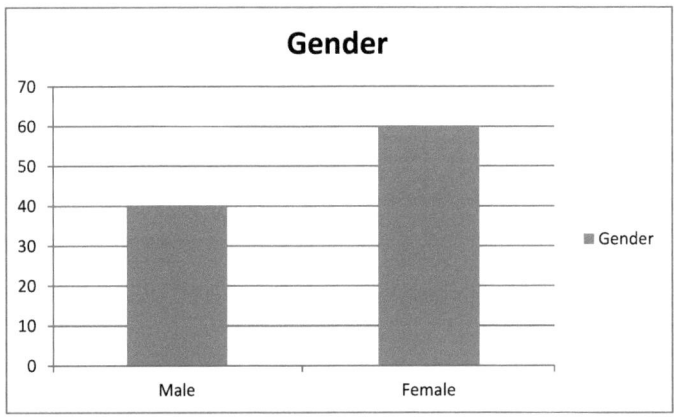

Table 3.b: Graphical display of Q2

For the question "what brand of shoes do you prefer?" 22 (22.00%) of the respondents answered Gibi, 15 (15.00%) respondents answered Russ, 32 (32.00%) of the respondents answered Hush Puppies, and 31 (31.00%) of the respondents answered others, in a total of 100 (100%) respondents which the majority of the respondents prefer Hush Puppies.

| SCHOOL SHOES | FREQUENCY | RELATIVE FRIQUENCY | PERCENTAGE |
|---|---|---|---|
| GIBI | 22 | 0.22 | 22% |
| RUSS | 15 | 0.15 | 15% |
| HUSG PUPPIES | 32 | 0.32 | 32% |
| OTHERS | 31 | 0.31 | 31% |
| TOTAL | 100 | 1.00 | 100% |

*Table 4.a: Frequency chart of Q4*

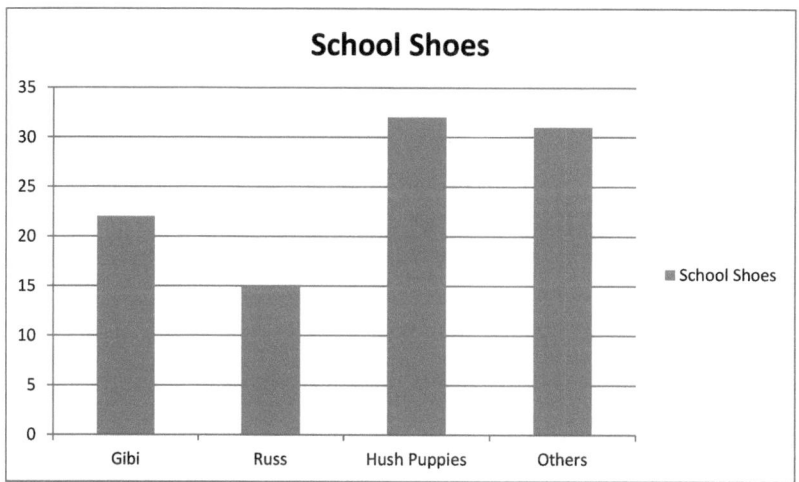

*Table 4.b: Graphical display of Q4*

In the question "rate your selected brand from 1-5" in comfort, 42 (42.42%) of the respondents rated excellent (1), 45 (45.45%) of the respondents rated very good (2), 12 (12.12%) rated good, and 0 (0.00%) of the respondents rated fair and poor with the majority preferring their brands to have excellent comfort.

| School shoes (comfort) | | | |
|---|---|---|---|
| 5 | 42 | 0.42 | 42% |
| 4 | 46 | 0.46 | 46% |
| 3 | 1 | 0.01 | 1% |
| 2 | 0 | 0 | 0% |
| 1 | 0 | 0 | 0% |

*Table 5.a : Frequency distribution of Q5-comfort*

In design 29 (29.29%) of the respondents rated excellent, 50 (50.51%) respondents rated very good, 20 (20.20%) of the respondents rated good, and 0 (0.00%) respondents rated fair and poor with the overall rating of their brands to have good design.

| School shoes (design) | | | |
|---|---|---|---|
| 5 | 29 | 0.29 | 29% |
| 4 | 51 | 0.51 | 51% |
| 3 | 20 | 0.20 | 20% |
| 2 | 0 | 0 | 0% |
| 1 | 0 | 0 | 0% |

*Table 5.b : Frequency distribution of Q5-design*

In durability 37 (37.37%) respondents rated their brands excellent, 48 (48.48%) respondents rated very good, 12 (12.12%) respondents rated good and 0 (0.00%) respondents rated fair and poor with the overall rating of their brand to have excellent durability.

| School shoes (durability) | | | |
|---|---|---|---|
| 5 | 37 | 0.37 | 37% |
| 4 | 49 | 0.49 | 49% |
| 3 | 12 | 0.12 | 12% |
| 2 | 1 | 0.01 | 1% |
| 1 | 1 | 0.01 | 1% |

*Table 5.c : Frequency distribution of Q5-durability*

In affordability 22 (22.22%) respondents rated excellent, 38 (38.38%) rated their brands as very good, 20 (20.20%) rated their brands as good, 1 (1.01%) rated their brands as fair and 1 (1.01%) rated their brands as poor, with an overall preference is for the brands to have a very good affordability or price range.

| School shoes (affordability) | | | |
|---|---|---|---|
| 5 | 22 | 0.22 | 22% |
| 4 | 39 | 0.39 | 39% |
| 3 | 20 | 0.20 | 20% |
| 2 | 17 | 0.17 | 17% |
| 1 | 2 | 0.02 | 2% |

*Table 5.d : Frequency distribution of Q5- affordability*

In brand impact 22 (22.22%) rated excellent, 39 (39.39%) rated very good, 28 (28.28%) rated good, 10 (10.10%) rated fair and 0 (0.00%) rated poor, with the overall preference is to have a brand with very good brand impact.

| School shoes (brand impact) | | | |
|---|---|---|---|
| 5 | 22 | 0.22 | 22% |
| 4 | 40 | 0.40 | 40% |
| 3 | 28 | 0.28 | 28% |
| 2 | 10 | 0.10 | 10% |
| 1 | 0 | 0 | 0% |

*Table 5.e: Frequency distribution of Q5- brand impact*

In overall quality 31 (31.96%) respondents rated excellent, 49 (49.48%) respondents rated very good, 18 (18.56%) respondents rated good and 0 (0.00%) respondents rated fair and poor, with the overall preference of having a good overall quality.

| School shoes (over-all quality) | | | |
|---|---|---|---|
| 5 | 32 | 0.32 | 32% |
| 4 | 49 | 0.49 | 49% |
| 3 | 19 | 0.19 | 19% |
| 2 | 0 | 0 | 0% |
| 1 | 0 | 0 | 0% |

*Table 5.f: Frequency distribution of Q5- overall quality*

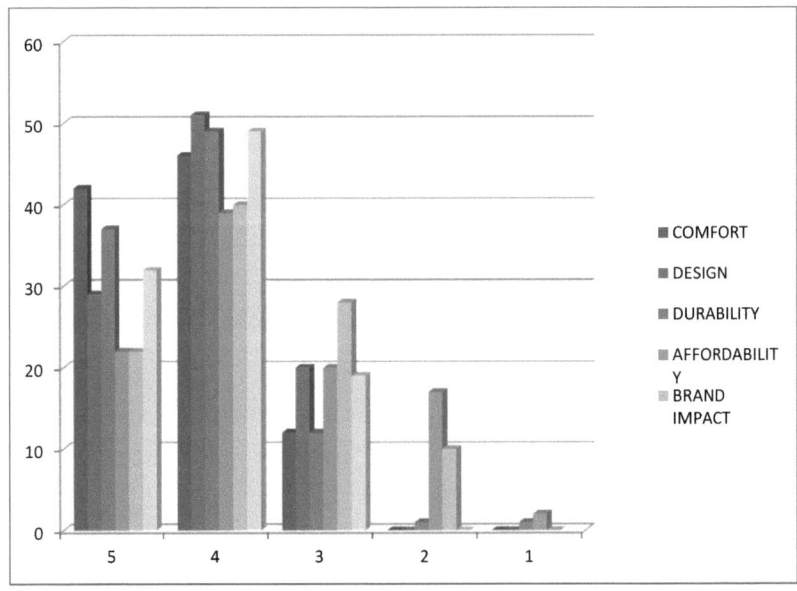

*Table 5.g: Graphical display of Q5*

For the question "what brand sneakers do you prefer?" 33 (33.00%) of the respondents answered Converse, 40 (40.00%) respondents answered Vans, 14 (14.00%) of the respondents answered Keds, and 12 (12.00%) of the respondents answered others, in a total of 100 (100%) respondents which the majority of the respondents prefer Vans.

| SNEAKERS | FREQUENCY | RELATIVE FREQUENCY | PERCENTAGE |
|----------|-----------|--------------------|------------|
| CONVERSE | 33 | 0.33 | 33% |
| VANS | 40 | 0.40 | 40% |
| KEDS | 14 | 0.14 | 14% |
| OTHERS | 12 | 0.12 | 12% |
| TOTAL | 100 | 1.00 | 100% |

*Table 6.a: Frequency distribution of Q6*

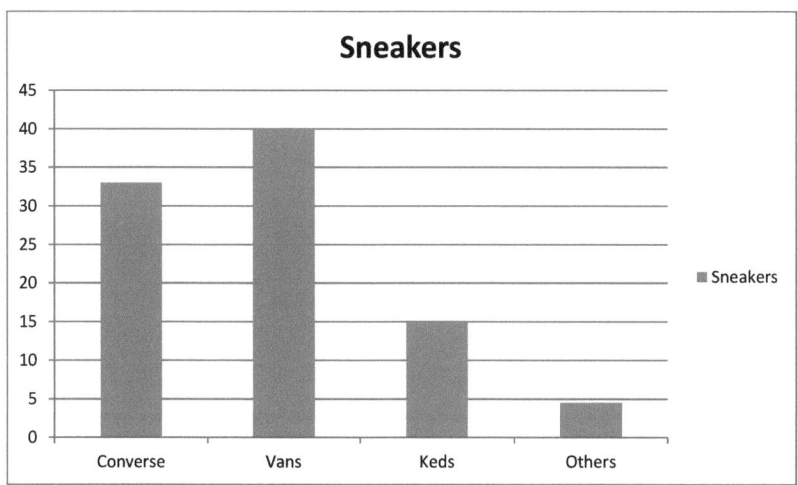

*Table 6.b: Graphical display of Q6*

In the question "rate your selected brand from 1-5" in comfort, 53 (53%) of the respondents rated excellent (1), 39 (39%) of the respondents rated very good (2), 8 (8%) rated good, and 0 (0.00%) of the respondents rated fair and poor with the majority preferring their brands to have excellent comfort.

| Sneakers (comfort) | | | |
|---|---|---|---|
| 5 | 53 | 0.53 | 53% |
| 4 | 39 | 0.39 | 39% |
| 3 | 8 | 0.08 | 8% |
| 2 | 0 | 0 | 0% |
| 1 | 0 | 0 | 0% |

*Table 7.a : Frequency distribution of Q7- comfort*

In design 64 (64%) of the respondents rated excellent, 29 (29%) respondents rated very good, 6 (6%) of the respondents rated good, 2 (2%) respondents rated fair and no respondent rated poor; with the overall rating of their brands to have an excellent design.

| Sneakers (design) | | | |
|---|---|---|---|
| 5 | 64 | 0.64 | 64% |
| 4 | 29 | 0.29 | 29% |

| | | | |
|---|---|---|---|
| 3 | 6 | 0.06 | 6% |
| 2 | 1 | 0.01 | 1% |
| 1 | 0 | 0 | 0% |

*Table 7.b : Frequency distribution of Q7- design*

In durability 51 (51%) respondents rated their brands excellent, 38 (38%) respondents rated very good, 9 (9%) respondents rated good and 2 (2%) respondents rated fair and no one rated poor; with the overall rating of their brand to have excellent durability.

| Sneakers (durability) | | | |
|---|---|---|---|
| 5 | 51 | 0.51 | 51% |
| 4 | 38 | 0.38 | 38% |
| 3 | 9 | 0.09 | 9% |
| 2 | 2 | 0.02 | 2% |
| 1 | 0 | 0 | 0% |

*Table 7.c : Frequency distribution of Q7- durability*

In affordability 17 (17%) respondents rated excellent, 40 (40%) rated their brands as very good, 32 (32%) rated their brands as good, 11 (11%) rated their brands as fair and no one rated their brands as poor, with an overall preference is for the brands to have an excellent affordability or price range.

| Sneakers (affordability) | | | |
|---|---|---|---|
| 5 | 17 | 0.17 | 17% |
| 4 | 40 | 0.40 | 40% |
| 3 | 32 | 0.32 | 32% |
| 2 | 11 | 0.11 | 11% |
| 1 | 0 | 0 | 0% |

*Table 7.d : Frequency distribution of Q7- affordability*

In brand impact 58 (58%) rated excellent, 33 (33%) rated very good, 9 (9%) rated good, no one rated fair and poor, with the overall preference is to have a brand with an excellent brand impact.

| Sneakers (brand impact) | | | |
|---|---|---|---|
| 5 | 58 | 0.58 | 58% |
| 4 | 33 | 0.33 | 33% |
| 3 | 9 | 0.09 | 9% |
| 2 | 0 | 0 | 0% |
| 1 | 0 | 0 | 0% |

*Table 7.e : Frequency distribution of Q7- brand impact*

In overall quality 53 (53%) respondents rated excellent, 37 (37%) respondents rated very good, 10 (10%) respondents rated good and 0 (0.00%) respondents rated fair and poor, with the overall preference of having an excellent overall quality.

| Sneakers (over-all quality) | | | |
|---|---|---|---|
| 5 | 53 | 0.53 | 53% |
| 4 | 37 | 0.37 | 37% |
| 3 | 10 | 0.10 | 10% |
| 2 | 0 | 0 | 0% |
| 1 | 0 | 0 | 0% |

*Table 7.f: Frequency distribution of Q7- overall quality*

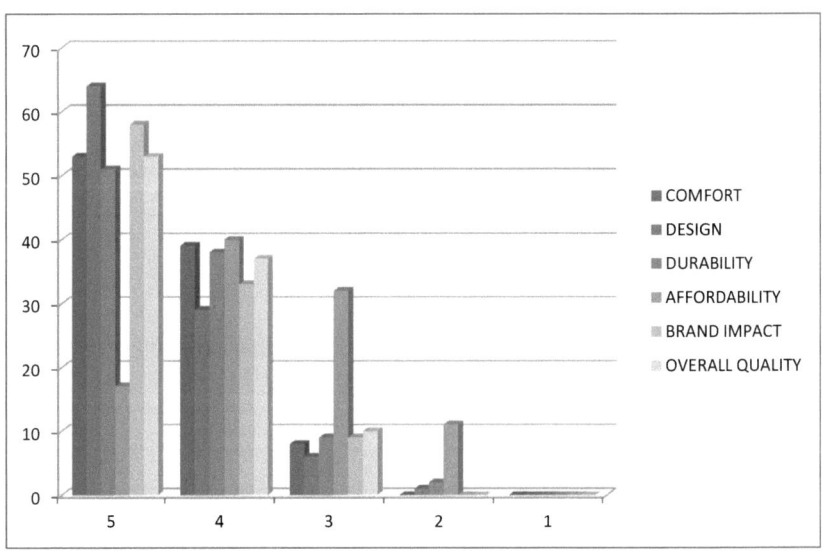

*Table 7.b: Graphical display of Q7*

For the question "what brand of rubber shoes do you prefer?" 72 (72.00%) of the respondents answered Nike, 23 (23.00%) respondents answered Adidas, 1 (1.00%) of the respondents answered Hush Puppies, and 4 (4.00%) of the respondents answered others, in a total of 100 (100%) respondents which the majority of the respondents prefer hush puppies.

| RUBBER SHOES | FREQUENCY | RELATIVE FREQUENCY | PERCENTAGE |
|---|---|---|---|
| NIKE | 72 | 0.72 | 72% |
| ADIDAS | 23 | 0.23 | 23% |
| ADVAN | 1 | 0.01 | 01% |
| OTHERS | 4 | 0.04 | 04% |
| TOTAL | 100 | 1.00 | 100% |

*Table 8.a : Frequency distribution of Q8*

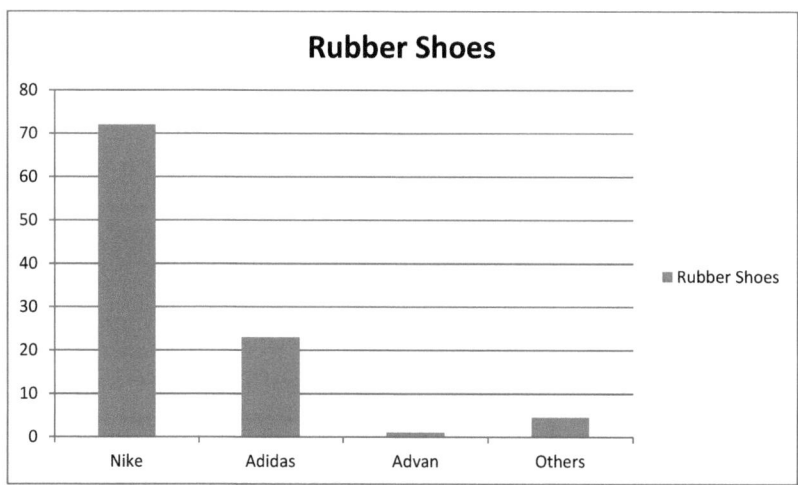

*Table 8.b: Graphical display of Q8*

In the question "rate your selected brand from 1-5" in comfort, 63 (63.00%) of the respondents rated excellent (1), 36 (36.00%) of the respondents rated very good (2), 1 (1.00%) rated good, and 0 (0.00%) of the respondents rated fair and poor with the majority preferring their brands to have excellent comfort.

| Rubber Shoes (comfort) | | | |
|---|---|---|---|
| 5 | 63 | 0.63 | 63% |
| 4 | 36 | 0.36 | 36% |
| 3 | 1 | 0.01 | 1% |
| 2 | 0 | 0 | 0% |
| 1 | 0 | 0 | 0% |

*Table 9.a : Frequency distribution of Q9- comfort*

In design 67 (67.00%) of the respondents rated excellent, 24 (24.00%) respondents rated very good, 9 (9.00%) of the respondents rated good, and 0 (0.00%) respondents rated fair and poor with the overall rating of their brands to have good design.

| Rubber Shoes (design) | | | |
|---|---|---|---|
| 5 | 67 | 0.67 | 67% |
| 4 | 24 | 0.24 | 24% |
| 3 | 9 | 0.09 | 9% |
| 2 | 0 | 0 | 0% |
| 1 | 0 | 0 | 0% |

*Table 9.b : Frequency distribution of Q9- design*

In durability 65 (65.00%) respondents rated their brands excellent, 29 (29.00%) respondents rated very good, 6 (6.00%) respondents rated good and 0 (0.00%) respondents rated fair and poor with the overall rating of their brand to have excellent durability.

| Rubber Shoes (durability) | | | |
|---|---|---|---|
| 5 | 65 | 0.65 | 65% |
| 4 | 29 | 0.29 | 29% |
| 3 | 6 | 0.06 | 6% |
| 2 | 0 | 0 | 0% |
| 1 | 0 | 0 | 0% |

*Table 9.c : Frequency distribution of Q9- durability*

In affordability 18 (18.00%) respondents rated excellent, 31 (31.00%) rated their brands as very good, 30 (30.00%) rated their brands as good, 19 (19.00%) rated their brands as fair and 2 (2.00%) rated their brands as poor, with an overall preference is for the brands to have a very good affordability or price range.

| Rubber Shoes (affordability) | | | |
|---|---|---|---|
| 5 | 18 | 0.18 | 18% |
| 4 | 31 | 0.31 | 31% |
| 3 | 30 | 0.30 | 30% |
| 2 | 19 | 0.19 | 19% |
| 1 | 2 | 0.02 | 2% |

*Table 9.d: Frequency distribution of Q9- affordability*

In brand impact 72 (72.73%) rated excellent, 25 (25.25%) rated very good, 2 (2.02%) rated good, 10 (10.10%) rated fair and 0 (0.00%) rated poor, with the overall preference is to have a brand with very good brand impact.

| Rubber Shoes (brand impact) | | | |
|---|---|---|---|
| 5 | 73 | 0.73 | 73% |
| 4 | 25 | 0.25 | 25% |
| 3 | 2 | 0.02 | 2% |
| 2 | 0 | 0 | 0% |
| 1 | 0 | 0 | 0% |

*Table 9.e: Frequency distribution of Q9- brand impact*

In overall quality 62 (62.00%) respondents rated excellent, 35 (35.00%) respondents rated very good, 3 (3.00%) respondents rated good and 0 (0.00%) respondents rated fair and poor, with the overall preference of having a good overall quality.

| Rubber Shoes (overall quality) | | | |
|---|---|---|---|
| 5 | 62 | 0.62 | 62% |
| 4 | 35 | 0.35 | 35% |
| 3 | 3 | 0.03 | 3% |
| 2 | 0 | 0 | 0% |
| 1 | 0 | 0 | 0% |

*Table 9.f : Frequency distribution of Q9- overall quality*

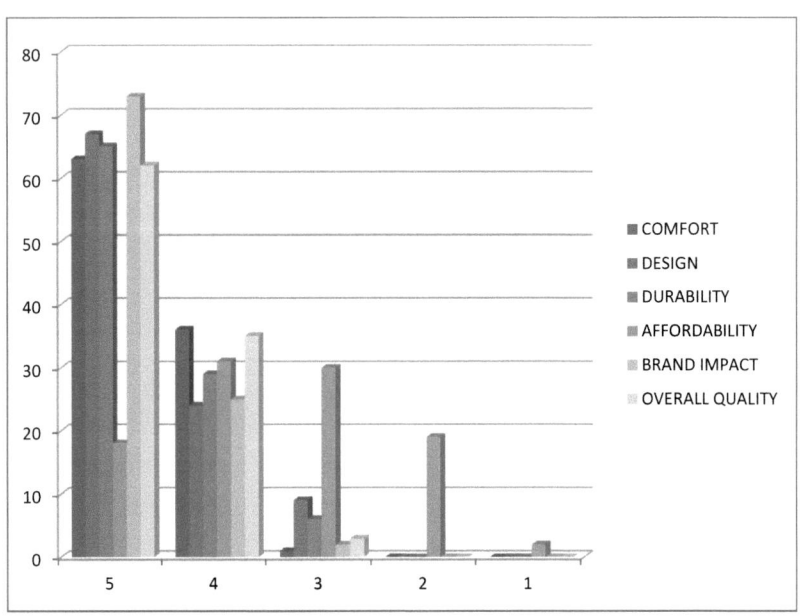

*Table 9.g: Graphical display of Q9*

CHAPTER 4: SUMMARY, CONCLUSION, RECOMMENDATION

SUMMARY OF FINDINGS

Out of a hundred who answered the survey, there were 2 out of 5 male students and 3 out of 5 who were female. Most of them being Advertising majors (64%), Painting Majors (15%), Industrial Design Majors (12%), then Interior Design Majors (9%).

According to the results found, when it comes to school shoes, most students would prefer the brand Hush Puppies (32%) rather than Gibi (22%), Russ (15%), or others (31%). Believing that their chosen brand has Very Good Comfort (46%), Very Good Design (51%), Very Good Durability (49%), Very Good Affordability (39%), Very Good Brand Impact (40%), and Very Good Overall Quality (49%).

In terms of Sneakers, students prefer the brand Vans (40%), rather than Converse (33%), Keds (15%), or Others (12%). Them stating that the brand they have chosen is Excellent on Comfort (53%), Excellent on Design (64%), Excellent on Durability (51%), Very Good on Affordability (39%), Excellent on Brand Impact (58%), and Excellent on Overall Quality (53%).

And when talking about Rubber Shoes, most students would prefer the brand Nike (72%) rather than Adidas (23%), Advan (1%), and Others (4%). For which they believe it having Excellent on Comfort (63%), Excellent on Design (67%), Excellent on Durability (65%), Very Good on Affordability (31%), Excellent on Brand Impact (72%), and Excellent on Overall Quality (62%).

From this results the researchers could then say that when it comes to their chosen brand, they could say that comfort, design, durability, brand impact, and overall quality could all reach excellent rating but it terms of affordability, very good is its highest rate according to the average. A good quality could mean an above average price.

CONCLUSION

The researchers came to the conclusion that for CFAD students prefer Hush Puppies for school shoes, Vans for sneakers and Nike for rubber shoes. These brands were selected due to the high amount of comfort, according to the survey as well.

RECOMMENDATION

The researchers recommend to the CFAD students that in choosing shoes, one should consider the quality of the products. The comfort and durability of the product is also important.

Based on the foregoing findings of the study, the following are recommended for future enhancement of the study regarding the Preference of CFAD students on selected brands of shoes.

1. The next researchers must conduct a survey with more than 100 respondents.

2. The next survey must be conducted both online and actual.

3. The next researchers should conduct the survey having equal number of male and female respondents, to give fair chances and equal response from both sexes.

BIBLIOGRAPHY

SurveyMonkey (2015, May 04). Retrieved from http://en.wikipedia.org/wiki/SurveyMonkey

**Consent Letter**

Dear respondent:

We are second year students of the College of Fine Arts and Design of the University of Santo Tomas and we are currently working on our research for our statistics course entitled, The Preference of CFAD Students on Different Brands of Shoes. As a CFAD student may we request for your participation in this research. It would be very helpful and it would provide us more information to complete this paper. Please find attached survey form.

Rest assured all information will be kept confidential.

We thank you in advance for your participation in our research.

The Researchers,
Cruz, Patricia
Guardiano, Clarice
Infante, Ruselle
Lacap, Dane
Lantin, Ivan
Muriel, Lyka

APPENDIX B

1. What brand of school shoes do you prefer?

    a. Gibi        b. Russ        c. Hush Puppies            d. Others:_____

Rate your selected brand from from 1-5 :

1- Excellent        2- Very Good        3-Good            4 – Fair
5 - Poor

|  | 1 | 2 | 3 | 4 | 5 |
|---|---|---|---|---|---|
| Comfort |  |  |  |  |  |
| Design |  |  |  |  |  |
| Durability |  |  |  |  |  |
| Affordability |  |  |  |  |  |
| Brand Impact |  |  |  |  |  |
| Overall Quality |  |  |  |  |  |

2. What brand of sneakers do you prefer?

    a. Converse    b. Vans        c. Keds d. Others:_____

Rate your selected brand from from 1-5 :

1- Excellent        2- Very Good        3-Good            4 – Fair
5 - Poor

|  | 1 | 2 | 3 | 4 | 5 |
|---|---|---|---|---|---|
| Comfort |  |  |  |  |  |
| Design |  |  |  |  |  |
| Durability |  |  |  |  |  |
| Affordability |  |  |  |  |  |
| Brand Impact |  |  |  |  |  |
| Overall Quality |  |  |  |  |  |

3. What brand of rubber shoes do you prefer?

    a. Nike       b. Adidas      c. Advan      d. Others:_____

    Rate your selected brand from from 1-5 :

    1- Excellent         2- Very Good       3-Good        4 – Fair

        5 - Poor

|  | 1 | 2 | 3 | 4 | 5 |
|---|---|---|---|---|---|
| **Comfort** |  |  |  |  |  |
| **Design** |  |  |  |  |  |
| **Durability** |  |  |  |  |  |
| **Affordability** |  |  |  |  |  |
| **Brand Impact** |  |  |  |  |  |
| **Overall Quality** |  |  |  |  |  |